BRITISH GEOLOG

CW00505125

Geology of the
Guildford district

a brief explanation of the geological map
Sheet 285 Guildford

R A Ellison
I T Williamson
A J Humpage

Bibliographic reference

ELLINSON, R A, WILLIAMSON, I T, and HUMPAGE, A J. 2002. Geology of the
Guildford district — a brief explanation of the geological map.
Sheet Explanation of the British Geological Survey.
1:50 000 Sheet 285 Guildford (England and Wales).

Keyworth, Nottingham: British Geological Survey

CONTENTS

Notes

The word 'district' refers to the area of Sheet 285 Guildford. National Grid references are given in square brackets prefixed by either SU or TQ denoting the 100 km square within which they fall. Lithostratigraphical symbols shown in brackets, for example (CaS) refer to symbols used on the 1:50 000 map. Borehole records referred to in the text are prefixed by the code of the National Grid 25 km² area within which the site falls, for example SU84SW, followed by the registration number in the BGS National Geosciences Records Centre.

Acknowledgements

The authors thank local authority officers of Surrey County Council and Guildford, Woking, Waverley and Mole Valley districts, and the Environment Agency regional office at Frimley for their ready co-operation in providing advice and information from their records. We acknowledge also the permission for access given by landowners in the district. We are particularly grateful to Drs C King and D Jolley for their work on the stratigraphy and palynology of the Mytchett boreholes.

The grid, where it is used on figures, is the National Grid taken from Ordnance Survey mapping.

Cover photograph
This sunken lane in the Hythe Formation was incised by the passage of hooved and wheeled transport before the road was metalled, and by erosion due to rain wash and wind [SU 998 425].
(*Photograph* C F Adkin, MN39732).

1   Introduction

This *Sheet Explanation* provides a summary of the geology of the district covered by the geological 1:50 000 Series Sheet 285 Guildford. It is written for the professional user and those who may have limited experience in the use of geological maps and may wish to be directed to further geological information about the district.

Descriptions of many important exposures formerly seen in the district can be found in Dines and Edmunds (1929). More recent information on these exposures together with newer sections, and accounts of important boreholes in the district are included in this account. The principal contributions towards understanding the geology of the district made since publication of the geological memoir concern the Hog's Back structure at the surface (Lake and Shephard-Thorn, 1985) and at depth (Smalley and Westbrook, 1982), and regional correlations of the Lower Greensand (Ruffell, 1992), the Chalk (Bristow et al., 1997), the lower Tertiary succession (Hester, 1965; Ellison, 1983; Ellison et al., 1994) and the London Clay (King, 1981). The depositional history of the higher level 'plateau' gravels in the east of the district was studied by Clarke and Dixon (1981) and Gibbard (1982) and river capture involving the Blackwater and Wey was explained by Worssam (1973). Other detailed information is available in BGS files and technical reports. This account follows revision mapping carried out at a scale of 1:10 000.

The Hog's Back is a narrow east–west-trending ridge of chalk reaching 152 m above OD that bisects the district. To the north are extensive pine woods and open heathland on gently rolling country with plateaux 100 to 120 m above OD underlain by Tertiary sands and clays. This area is dissected in the west by the River Blackwater valley, where there have been extensive gravel workings, and in the east

by the River Wey. South of the Hog's Back, the subdued topography is underlain mainly by sands of the Folkestone and Sandgate formations. The highest ground hereabouts lies east of Godalming where a wooded, deeply dissected area underlain by sandstone of the Hythe Formation rises to an escarpment crest locally higher than 200 m above sea level.

The oldest rocks in the district, proved in boreholes, are Ordovician and Silurian, marine shales and mudstones. These were laid down on a relatively stable area that was gently folded in Mid Devonian times during development of the main Caledonian fold belts in Wales and eastern England. Succeeding strata of Late Devonian to Carboniferous age were subject to folding, faulting and erosion during the Variscan orogeny and are only thinly preserved in the district. The frontal orogenic thrust within the Palaeozoic rocks dips to the south and trends roughly east–west through the northern part of the district. By the end of this orogeny, some 290 million years ago, the district lay across the southern edge of the London Platform, a relatively stable block that extended to Belgium.

Permian and Triassic sediments, laid down in arid conditions, may have been thinly deposited in this district. During late Triassic times and continuing until the early Cretaceous, crustal extension led to growth faulting and rift development on re-activated Variscan faults at the southern margin of the London Platform. These movements led to the formation of the Weald Basin, part of which lies in the southern part of the district. The basin fill consists of Jurassic marine shales with some limestone and sandstone (Penarth Group to Purbeck Group), and a Lower Cretaceous brackish to freshwater clastic sequence (Hastings Beds and Weald Clay). The London Platform was progressively overlapped by Jurassic sediments and was finally completely covered by the Oxford Clay, about 155 million years ago. A reduction in tectonic subsidence and gentle uplift, in early Cretaceous times, led to the Late Cimmerian unconformity and erosion on the London Platform. Regional post-rift subsidence then occurred causing further inundation of the London Platform and deposition of shallow marine sandstones and mudstones of the Lower Greensand and Gault. By late Cretaceous times, much of Britain lay beneath a sea with little terrigenous input in which the chalk was deposited.

Late Cretaceous regional uplift and tilting led to significant erosion of the Upper Chalk. The succeeding Tertiary sediments were laid down in shallow marine, brackish and lacustrine environments at the margin of a new depositional basin that extended westwards from the North Sea. At the same time, an important tectonic change was initiated leading to compressive inversion and reversal of the faults bounding the Weald Basin. This ultimately gave rise to gentle folding of the synclinal London Basin and uplift of the Weald Basin, culminating in the Miocene, by which time the Hog's Back Monocline had formed.

In Pliocene or early Quaternary time a minor marine incursion led to deposition of the Netley Heath Beds. Later in the Quaternary, the district was drained by rivers flowing north from the Chalk uplands and the Weald, ultimately to join an ancestral Thames that flowed across southern East Anglia to the North Sea. The river deposits are preserved as high-level terrace gravels in the west of the district. Head gravel deposits, and the intensely cryoturbated Headley Heath Beds and Clay-with-flints provide evidence of several periods of periglaciation that led to the solifluction of older river terrace deposits and unconsolidated Tertiary and Cretaceous sediments.

The Thames was diverted to its present course about 0.5 Ma by the southward progress of the Anglian ice sheet that halted to the north of this district. Following this major glacial episode the present valleys of the rivers Wey and

Blackwater became established as distinct drainage systems. In the period between the Anglian and the end of the Devensian stage, about 10 000 years ago, the Thames, Wey and Blackwater rivers laid down braided sand and gravel deposits. These are arranged in a series of river terrace deposits whose relative surface elevations became successively lower with time. The only change in the drainage pattern in this period occurred about 36 000 years ago when the headwaters of the River Blackwater were captured by the River Wey. Since the Devensian, the climate has gradually warmed and the rivers have deposited relatively fine-grained sediment and established meandering courses across the floodplains (Plate 1).

Plate 1   Flood plain gravels of the River Blackwater [SU 885 560] (A 12727).

2   Geological description

The concealed geology of the district, inferred from deep boreholes (Figure 1), seismic data and regional gravity and magnetic studies, is illustrated on the horizontal cross-section on the geological map.

Palaeozoic

The oldest rocks proved in the central part of the district are of Lower Palaeozoic age (see Figure 1) and lie within the Variscan fold belt sequence. Arenig (**Ordovician**) strata comprising 81 m of sandstones, sandy limestones and muddy siltstones with a 50 to 60° dip were proved in the Strat-A1 Borehole; fossiliferous, purple and grey-green mudstones 26.5 m thick, of Llandovery (**Silurian**) age, were proved in the Shalford Borehole. The basement intersected by the Coxbridge Borehole comprises mudstones also thought to be of Lower Palaeozoic age. Some of the sequence at the base of the Normandy Borehole was once considered to be Triassic or Permo–Triassic (Penarth and Mercia Mudstone groups) but may be Devonian in age.

At depth in the north of the district, on the London Platform, there probably occur mainly red sandstones and mudstones of Devonian age, similar to those under the adjacent Windsor (Sheet 269) district to the north and greater London (see Sumbler, 1996 for a review), although there are no deep boreholes.

Strata of Carboniferous age are thin and only preserved locally. Probable **Carboniferous** strata, represented by 26 m of off-white, bioclastic limestones with interbedded claystones, are proved at the base of the Albury Borehole. A 77.4 m-thick sequence of variegated mudstones and limestone breccia overlying the Lower Palaeozoic basement in the Shalford Borehole is either Carboniferous or Triassic in age.

Mesozoic

The most stratigraphically complete and thickest (about 2500 m) sequence occurs in the south of the district.

The **Mercia Mudstone (MMG)** and **Penarth (PnG)** groups of **Permo–Triassic** age comprise reddish brown and variegated mudstones and sandstones. They are 13.7 m and 34 m thick, and occur in the Shalford and Normandy boreholes, respectively. In the Coxbridge Borehole, strata of Triassic age are 69 m thick, and comprise 35 m of limestone and sandstone (Penarth Group) overlying claystones (Mercia Mudstone Group). **Jurassic** strata beneath the southern half of the district include the **Lias Group (Li)** to the **Purbeck Limestone Group (Pb)**. They are a maximum of 1600 m thick in the Weald Basin, attenuating markedly across the margin of the London Platform to 50 to 100 m because of stratigraphical onlap and also erosion of the younger Jurassic strata (see section on the 1:50 000 scale map). The overlying **Hastings Beds (HB), Weald Clay (WC)** and **Atherfield Clay (AC)** also thin northwards due to uplift of the London Platform leading to the Late Cimmerian unconformity.

Weald Clay Formation (WC)

Distribution: outcrops are in the core of the Peasmarsh Anticline [SU 985 465] and in the south-eastern corner of the district eastward from Rushett Common [TQ 025 420]. They are generally in low-lying ground, with low hills capped by residual superficial deposits or thin, relatively resistant beds. At depth, in the vicinity of the Hog's Back, the Weald Clay is overstepped by the Lower Greensand.

Lithology: clay and silty clay, interlaminated with clayey silt. At depth the sediments are pale to dark grey, greenish grey and red,

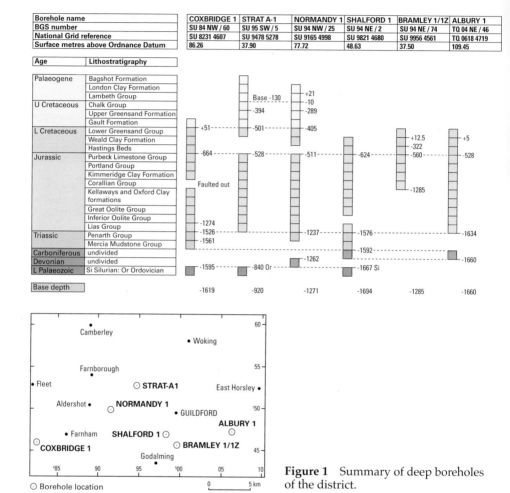

Borehole name		COXBRIDGE 1	STRAT A-1	NORMANDY 1	SHALFORD 1	BRAMLEY 1/1Z	ALBURY 1
BGS number		SU 84 NW / 60	SU 95 SW / 5	SU 94 NW / 25	SU 94 NE / 2	SU 94 NE / 74	TQ 04 NE / 46
National Grid reference		SU 8231 4607	SU 9478 5278	SU 9165 4998	SU 9821 4680	SU 9956 4561	TQ 0618 4719
Surface metres above Ordnance Datum		86.26	37.90	77.72	48.63	37.50	109.45

Age	Lithostratigraphy
Palaeogene	Bagshot Formation
	London Clay Formation
	Lambeth Group
U Cretaceous	Chalk Group
	Upper Greensand Formation
	Gault Formation
L Cretaceous	Lower Greensand Group
	Weald Clay Formation
	Hastings Beds
Jurassic	Purbeck Limestone Group
	Portland Group
	Kimmeridge Clay Formation
	Corallian Group
	Kellaways and Oxford Clay formations
	Great Oolite Group
	Inferior Oolite Group
	Lias Group
Triassic	Penarth Group
	Mercia Mudstone Group
Carboniferous	undivided
Devonian	undivided
L Palaeozoic	Si Silurian: Or Ordovician

Figure 1 Summary of deep boreholes of the district.

weathering in the top 5 to 7 m below the surface to shades of yellow and brown and pale grey-brown, but the primary red colour may persist. Several beds, mostly up to 5 m thick, of fine- to medium-grained, flaggy sandstones, and limestones with a characteristic fauna of freshwater gastropods ('*Paludina*' limestones) occur below the top 80 m that are exposed hereabouts (Thurrell et al., 1968; Worssam, 1978). Beds of dominantly red clay also form markers at several horizons, but none has been mapped in this district. Horizons with abundant clay ironstones, generally up to 10 cm in diameter occur at irregular intervals.

Thickness: up to a maximum of 454 m proved in boreholes; absent in the north.

LOWER GREENSAND GROUP (LGS)

The formations of the Lower Greensand Group (Figures 2, 3) thin progressively northwards, and at depth here only the Folkestone Formation and possibly the Atherfield Clay are present.

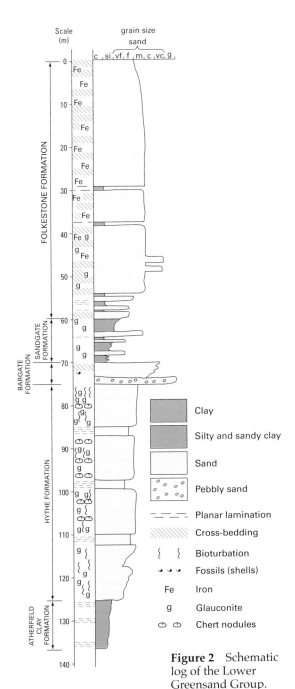

Figure 2 Schematic log of the Lower Greensand Group.

Key:

- Clay
- Silty and sandy clay
- Sand
- Pebbly sand
- Planar lamination
- Cross-bedding
- { } Bioturbation
- Fossils (shells)
- Fe Iron
- g Glauconite
- Chert nodules

Atherfield Clay Formation (AC)

Distribution: on the lower slopes of the Lower Greensand escarpment around Rushett Common [TQ 020 425], in low-lying areas in the Peasmarsh Anticline [TQ 010 470], and in two small inliers at Thorncombe Street [TQ 0018 4295] and east of Shamley Green [TQ 0440 4360]. It probably also occurs at depth across the north of the district.

Lithology: mainly clay and silty clay with subordinate sandy clay, in shades of blue, brown and yellow. Horizons with clay ironstone concretions up to 300 mm in size are common towards the base. The formation contains a rich molluscan fauna (Casey, 1961), which includes the ammonite *Prodeshayesites obsoletus*, indicative of the *fissicostatus* Zone (Woods, 1998), and corals and shells are recorded in some of the concretions (Simpson, 1985).

Thickness: from about 7 m at Rushett Common to 20 m in the Peasmarsh area.

Hythe Formation (Hy)

Distribution: in much of the south-eastern corner of the district. In the west, it is less extensive, and around Godalming outcrops are largely confined to the valleys. Not present at depth in some areas.

Lithology: typically pale grey, well-sorted, fine to medium-grained sand and weakly cemented sandstone, clayey in the lowest 15 m. The sands are predominantly quartzose with, particularly at the top, glauconite grains of fine to medium sand grade that give a distinctive speckled 'salt and pepper' appearance. The sediments are in general bioturbated and lamination is rare;

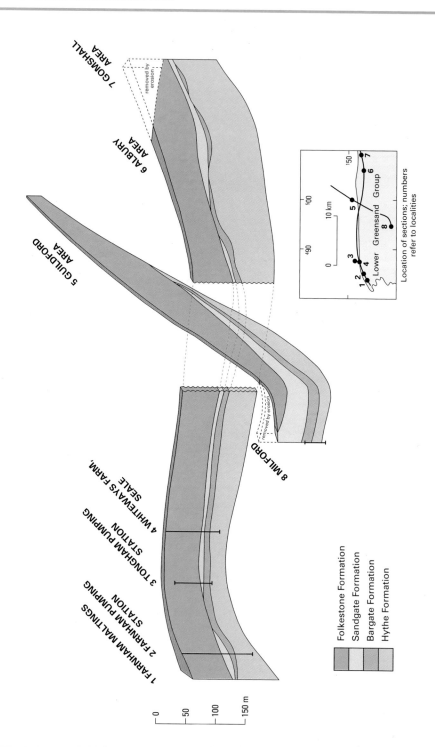

Figure 3   Variation of the Lower Greensand Group.

some cross-bedding is discernible in the uppermost units. In the middle part, chert nodules occur in beds of grey-brown, well-jointed, fine-grained sandstones up to 2 m thick. Relatively resistant beds of fine-grained sandstone, exposed in small sections [TQ 0445 4381; TQ 0208 4450], form positive features that have been mapped to the east and north of Shamley Green [TQ 0370 4420; TQ 0320 4455]. Three of them, each about 2 m thick, occur at about 10 m intervals, the highest being about 10 m below the top of the formation (Figure 2). Ruffell (1992b), in a regional review, identified three 'event' horizons, delineated by smectite-rich clays, phosphatic nodules or shell beds that may be a result of hiatuses in sedimentation during transgression. The main exposures are in the top part of the succession (10 m thick) at Lower Easing [SU 9468 4368] and in a quarry at Mill Hollow [TQ 0795 4253].

Thickness: the maximum is estimated as about 85 m, around Peaslake [TQ 08 44]; on the evidence of boreholes, only about 30 m are present south of Milford. There is no evidence for the exceptional 100 m thickness of Hythe Formation reported around Elstead by Ruffell (1992).

Bargate Formation (Bt)

Distribution: discontinuous crops occur south of the Hog's Back and around Albury Heath [TQ 058 466], Blackheath [TQ 038 464] and Farley Heath [TQ 046 430]. The formation forms a dissected plateau more than 8 km wide to the south and west of Godalming [SU 95 43] and there are small outliers around Shere [TQ 0780 4730] and Abinger [TQ 1020 4680]. The formation is more extensive in the east than previously recorded, particularly in the Albury area, on Blackheath, and around the southern part of Farley Heath.

Lithology: buff to brown, gritty, ferruginous cemented, cross-bedded, coarse-grained, with some beds of pale grey calcareous sandstone. At the base there is a local pebble bed that contains reworked clasts of Hythe Formation, glauconite pellets, quartz granules, brown phosphatic pebbles up to 10 mm in size, and pebbles of fossiliferous limestone of Jurassic age (derived from sediments on the London Platform to the north) (Meyer, 1868). Elsewhere the basal contact with the Hythe Formation is gradational, suggesting an element of recycling (Lake and Shephard-Thorn, 1985). The type section is at Blackheath Lane, Albury [TQ 0483 4700] where cross-bedding, in 15 m-wide channels, is exposed in up to 12 m of poorly sorted, calcareous cemented, coarse-grained sandstone. The succession is decalcified in other places, for example 550 m to the east in an exposure in Warren Lane [TQ 0536 4738] where there are about 6 m of cross-bedded, loose textured, coarse-grained sand with mainly subangular grains (Lake and Shephard-Thorn, 1985; Kirkaldy, 1933).

In the western part of the district, the best exposure of the Bargate Formation is at the top of the Milford bypass road cutting [SU 9435 4413]. This exposes up to 7 m of cross-bedded friable sandstone, weakly calcareous in places, with chert beds up to 0.3 m thick at 1 to 2 m intervals, and a prominent 0.5 m-thick weakly calcareous chert bed at the bottom. These beds give rise to a well-defined dip slope [SU 949 433] inclined at about 1 to 2° to the southwest. The basal beds are exposed in a road cutting at Eashing [SU 9483 4357].

Thickness: typically up to 10 m at outcrop around Godalming; 17 m were proved in boreholes in the west of the district and up to 20 m in boreholes on Crooksbury Common [SU 880 460]; it is absent locally east of Godalming and probably in the west around Farnham.

Sandgate Formation (SaB)

Distribution: crops out mainly around Elstead [SU 91 43], and in outliers adjacent to Charterhouse School [SU 960 455], on Munstead Heath [SU 982 427] and north and west of Peaslake [TQ 090 445]. Though it mostly overlies the Bargate Formation, as at Albury, in other locations the Bargate Formation is present where the Sandgate

Formation is not, and vice versa. Locally the two formations may be laterally equivalent.

Lithology: previously known as 'loamy beds' within the Folkestone Formation Dines and Edmunds, 1929 comprises loose, argillaceous, medium grey, glauconitic, ferruginous, fine- to coarse-grained sand and sandstone which is characteristically poorly sorted. Scattered lenses containing quartz granules and subangular clay pellets occur locally. The outcrop is typified by scrubby and locally wet heathland.

Thickness: very variable; around Albury [TQ 0480 4730] there are 5 m or less; at Church Croft [SU 9220 4660], 3 km north of Elstead, up to 45 m. The formation is absent at the foot of the Hog's Back escarpment between Compton [SU 9645 4774] and Shere [TQ 066 479].

Folkestone Formation (Fo)

Distribution: wide crops in the south-west and south-east, with a narrow belt between; characterised by rolling convex slopes with a sandy heathland dominated by birch, gorse and pine plantations. A poorly defined break of slope and a minor spring line locally marks the base of the formation. The formation is present at depth north of the Hog's Back, but in the far north may be overstepped by the Gault.

Lithology: predominantly loosely consolidated, ferruginous, medium-grained quartz sand, green-grey and glauconitic at depth and weathering to orange-brown. The basal beds of the formation locally include coarse-grained sand. Cross-bedding, in sets up to 2 m thick, occurs throughout. Beds of pale grey-green silt and clay generally 1 to 2 cm thick occur at intervals, particularly in the middle of the succession.

Vertical irregular-shaped veins and horizontal seams of iron oxide-cemented sandstone typically 5 to 10 cm wide, known as 'carstone', occur throughout along with similarly lined sand-filled tubes and cylindical concretionary nodules. Fragments of this are common in soils on the outcrop. Around Farnham, an argillaceous iron pan near the top of the formation is characterised by the ammonite *Farnhamia farnhamensis* (Casey, 1961; Owen, 1984). Thin iron-pan horizons, probably of Quaternary age, developed subparallel to the present ground surface, are common on Blackheath and Farley Heath, where they locally impede drainage and create boggy areas in topographic lows. The best exposure is in Albury Sand Pit [TQ 057 483] where about 20 m of Folkestone Formation show good examples of 'carstone' development and large foresets in the sands that were deposited by currents flowing from the west (Plate 2). A near-surface ferruginous layer up to 2 m thick is exposed in a roadside section at Waverley Lane, Compton [SU 8612 4610]. Up to 10 m of Folkestone Formation were noted at Gomshall Station [TQ 089 479].

Thickness: in the northernmost crop, along the foot of the Chalk escarpment, it varies from 32 to 60 m; in the south-west there are

Plate 2 Cross-bedding in the Folkestone Formation, Albury Sand pit, Weston Wood [TQ 055 484] (A 13922).

an estimated 50 to 80 m. This significant variation may be caused by syndepositional tectonic activity (Ruffell, 1992). At depth the formation thins northwards, as shown by the 44 m and 22 m proved in the Normandy and Strat-A1 boreholes respectively (Figure 1), and illustrated on the 1:50 000 map cross-section.

Gault Formation (G)

Distribution: outcrops are in the south-west of the district at Rowledge [SU 82 43] and a narrow strip along the foot of the Hog's Back. Gault is also present at depth north of the Hog's Back.

Lithology: typically dark grey to bluish grey mudstone that weathers to brown and mottled pale grey and purplish grey stiff clay. The sediments are locally micaceous and contain scattered pyrite nodules and horizons with septarian nodules and ferruginous concretions. The basal bed, about one metre thick, is markedly glauconitic and sandy with sporadic fossiliferous phosphate nodules. Phosphatic nodules are recorded near the top of the formation near Farnham and in several beds in the basal 5 m in a pit at Wrecclesham (Dines and Edmunds, 1929). Other horizons may occur at intervals through the succession. The nodules probably formed during periods of winnowing of unconsolidated material on the sea bed (Knight, 1999) during short depositional hiatuses.

The Gault has been divided (Forster et al., 1995) into a lower unit of medium and dark grey, variably silty mudstones, dominated by illite and kaolinite, and an upper unit that is paler grey, has a higher calcium carbonate content and in which smectite is the dominant clay mineral. These units probably coincide with the 'Upper Gault' and 'Lower Gault' recognised largely on palaeontological evidence by Owen (1975). Owen (1992) also detailed the nature of the Gault–Lower Greensand junction from localities in this district.

Thickness: the thickest sequence proved is 122 m in a borehole at Tongham (SU84NE66 [SU 8776 4840]). Some 109 m were proved in the Normandy Borehole. The thickness at outcrop is generally about 80 to 93 m, but is 50 m or less along the Hog's Back, probably due to a combination of structural control and internal shear caused by tectonic compression. This is supported by an exposure at Runfold [SU 865 476] which showed slickensided and polished clays enveloping irregular sand bodies (Lake and Shephard-Thorn, 1991). In regional terms, the 'Lower Gault' thins northwards but the 'Upper Gault' is likely to be of more constant thickness.

Upper Greensand Formation (UGS)

Distribution: crops out on a low positive feature close to the base of the main Hog's Back escarpment and along the lower slopes of the Chalk escarpment of the North Downs. Its distribution at depth in the north is not known, but it appears not to be present in the Normandy Borehole.

Lithology: the base is gradational with the underlying Gault Formation and is taken at the incoming of pale to medium grey siltstones, which weather to low density, loose sediments known as 'hearthstone'. The beds are typically bioturbated, and near the top of the formation, there is a massive, hard, grey-white, siliceous sandstone known as 'firestone' that was dug locally. There are small exposures in former quarries south of Guildford [for example at SU 9790 4829; SU 9727 4825].

Thickness: varies from about 12 to 30 m at outcrop, generally thickening eastwards at the expense of the underlying Gault (Lake and Shephard-Thorn, 1985).

CHALK GROUP (Ck)

The Chalk was traditionally divided into Upper, Middle and Lower units defined by biostratigraphical zonation and gross lithological characteristics. Bristow et al. (1997) produced a formal lithostratigraphy that defined the three units as formations and slightly modified their boundaries (Figure 4). This scheme has been followed in the resurvey of the district, but a delineation of members in the Chalk has only been carried out in the Upper Chalk (Figure 5), mainly

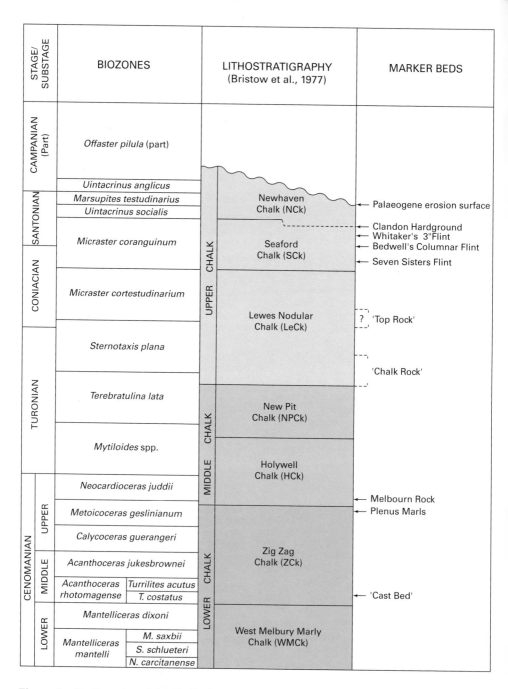

Figure 4   Stratigraphy of the Chalk Group.

because of the relatively narrow crops of Middle and Lower Chalk. Identification of these members has been aided by the assessment of fossil collections from former pits in the Upper Chalk (Woods, 1997, 1998; John, 1980) (Figure 5). Further details of these and other chalk pits are given in Dines and Edmunds (1929), Lake and Shephard-Thorn (1991), and Shephard-Thorn and Lake (1991).

Distribution: there is continuous outcrop from near Farnham in the west to near Gomshall in the east.

Thickness: the thickness of the group varies between 240 and 280 m.

Lower Chalk Formation (LCk)

Lithology: pale to medium grey, argillaceous, micritic limestones that in places form a subsidiary escarpment at the foot of the North Downs. The lower part is typically thinly bedded, with rhythmic alternations of grey to blue-grey, clay-rich beds and impure limestones (**West Melbury Marly Chalk** of Bristow et al., 1997). It passes upward into more massively bedded, smoother textured chalk with thinner marls, the **Zig Zag Chalk** of Bristow et al. (1997). Limestones, characterised by the inoceramid bivalve *Inoceramus tenuis*, and overlain by a silty chalk horizon with a distinctive brachiopod and bivalve fauna (the 'Cast Bed'), are distinctive markers at the base of this member. Clay-rich beds, the **Plenus Marls**, mark the top of the Lower Chalk.

Thickness: 75 m thick, west of Guildford, possibly thinning eastwards to 65 m near Dorking, 6 to 7 km east of this district (Lake and Shephard-Thorn, 1985). Borehole data however shows a greater range with a minimum of 43 m at East Horsley (borehole TQ 05SE3 [TQ 0981 5306]) and a maximum of 84 m around West Clandon (borehole TQ 05SW127 [TQ 0472 5064]).

Middle Chalk Formation (MCk)

Lithology: white chalk with few flints. The base is typically a hard, nodular feature-forming bed called the **Melbourn Rock**. This is overlain by hard, nodular chalk

with thin marls, becoming shell-rich in the higher part. Together these strata comprise the **Holywell Chalk** of Bristow et al. (1997). Higher beds are typically non-nodular, softer, smoother textured and massively bedded chalk with regularly developed thin marl beds and these equate to the **New Pit Chalk** of Bristow et al. (1997).

Thickness: generally about 65 to 75 m in the Guildford and Dorking areas. The minimum recorded is 41 m at East Horsley (borehole TQ 05SE3).

Upper Chalk Formation (UCk)

The base of the Upper Chalk is taken at the base of the hard, nodular chalk known as the **Lewes Nodular Chalk** (Bristow et al., 1997). This boundary (which lies in the upper part of the *T. lata* Zone) is in general 1 to 5 m lower than the traditional base of the Upper Chalk that was drawn at the base of the *S. plana* Zone (see Figure 4).

Lithology: the lowest part of the succession is characterised by beds of indurated chalk (hardgrounds) and marl seams (Lewes Nodular Chalk). The stratigraphically lowest occurrence of conspicuous flint bands lies in these beds. The Lewes Nodular Chalk is overlain by the **Seaford Chalk**, a relatively soft, thickly bedded, flinty chalk with thin marl seams in its lower part. The Clandon Hardground, indurated chalk with a glauconitised top surface developed on a 30 cm-thick, hard, limonite-stained bed of 'chalkstone' (Robinson, 1986), marks the top of the Seaford Chalk. Above it the **Newhaven Chalk** is relatively soft, thick to massively bedded chalk with marl seams and bands of scattered nodular flints. The youngest Chalk exposed in the district lies in the middle of the Newhaven Chalk. It was probably exposed in temporary sections at Guildford Park [around SU 98 49] (Edmunds, 1927; Dines and Edmunds, 1929) and is described by Woods (1997, 1998) from the same general area west of the Royal Surrey County Hospital, Guildford.

Thickness: estimated by Lake and Shephard-Thorn (1985) as generally about 120 to 125 m

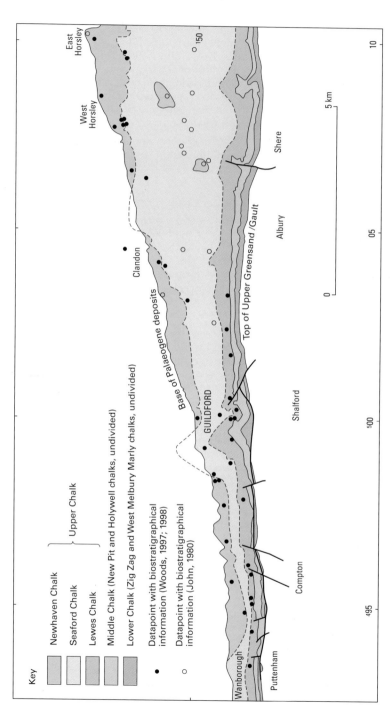

Figure 5   Map showing subdivision of the Upper Chalk.

Key

Newhaven Chalk ⎱
Seaford Chalk ⎰ Upper Chalk
Lewes Chalk

Middle Chalk (New Pit and Holywell chalks, undivided)

Lower Chalk (Zig Zag and West Melbury Marly chalks, undivided)

● Datapoint with biostratigraphical information (Woods, 1997; 1998)

○ Datapoint with biostratigraphical information (John, 1980)

East Horsley
West Horsley
Clandon
Base of Palaeogene deposits
GUILDFORD
Top of Upper Greensand /Gault
Shere
Albury
Shalford
Compton
Wanborough
Puttenham

150

0   5 km

10
05
500
495

at outcrop. A revaluation of data in the eastern part of the district at East Horsley suggests up to 164 m (borehole TQ05SE3) may be present.

Palaeogene

Thanet Sand Formation (TS)

Distribution: in the central and eastern parts of the London Basin, the Thanet Sand is the oldest Palaeogene formation. It is present at depth in the north-east of this district but absent at crop west of the River Wey and at depth in the west of the district.

Lithology: silty, fine-grained sand, typically green or grey when fresh, weathering pale ochreous brown. The basal bed (the Bullhead Bed) is up to 0.5 m thick and consists of nodular and rounded flint-gravel in a clayey sand matrix. Dines and Edmunds (1929, pp.81–82) described water-well sections and exposures. There are no current exposures.

Thickness: up to 6 m, the maximum being proved in the north-west suburbs of Guildford.

LAMBETH GROUP (LMB)

This group comprises the Woolwich and Reading Beds of previous accounts and is divided into the Upnor Formation (UPR) at the base overlain by the Woolwich (WL) and Reading (RB) formations (Ellison et al., 1994). During the resurvey of the district it was not practicable to map out the individual formations, and most borehole logs are insufficiently detailed to identify them. However, the general lithology of each formation is known from borehole cores and exposures and is described below.

Distribution: the Lambeth Group overlies the Thanet Sand in the east, overstepping on to the Chalk westwards at about the longitude of Woking. In addition to the main crop from Farnham to East Horsley, there are outliers on the Chalk dip slope around Hook Wood [TQ 077 506] and Barnet Wood [TQ 074 500]. At depth the group is ubiquitous beneath the London Clay. The boundary between the Chalk and the Lambeth Group is marked by a series of prominent sinkholes in valleys in the Farnham Park area.

Lithology: the **Upnor Formation**, previously known as the 'Bottom Bed' consists of green-grey, glauconitic, sandy clays, generally with local, thin, flint pebble beds and lenses of pebbles, particularly at the base. The basal pebble bed may be lithologically indistinguishable from the Bullhead Bed at the base of the Thanet Sand. The thickness is generally less than 2 m. The **Reading Formation** comprises much of the Lambeth Group. It consists mainly of stiff to firm, mottled red, brown, pale grey and blue-grey clay with subordinate, relatively thin beds of loose fine-grained sands and thinly laminated silt. In the Windsor and Bracknell district to the north (Ellison and Williamson, 1999) these sands, interpreted as major channel sand bodies, make up a significant proportion of the formation. Similar deposits in this district have been observed only in a ditch section to the south of Poyle Farm [SU 8999 4861]. The **Woolwich Formation**, probably only up to 3 m thick, occurs locally at the top of the Lambeth Group. It consists of dark grey, shelly clay and lignitic clay and was recorded in former brickpits around Guildford (Dines and Edmunds, 1929, p.84). Lignitic beds were seen recently in temporary excavations at Manor Farm [SU 8874 4843].

Thickness: boreholes generally prove between 17 and 25 m, but there is no systematic variation. The minimum recorded is 15 m near Guildford and the maximum 39 m at Ockham, but both these records, especially the latter, may be inaccurate.

THAMES GROUP (TGp)

Harwich Formation (Har)

Distribution: comprises sand-dominated beds between the **Lambeth Group** and the **London Clay Formation** (Ellison et al., 1994) that include strata described by Dines and Edmunds (1929) as the 'Basement Bed of the London Clay'. The formation is inconsistently developed and generally insufficiently thick to show on the geological map but is recognised in exposures and borehole core.

Lithology: silty, fine-grained sand and sandy clay, greenish brown, locally shelly. The strata were formerly exposed near Guildford and proved in wells around Aldershot.

Thickness: 0.6 to 3.7 m (in borehole logs).

London Clay Formation (LC)

Distribution: crops out, and is present at depth, north of a line between north-west Farnham and East Horsley. It gives rise to relatively low-lying, undulating ground, generally characterised by heavy clay soils.

Lithology: consists mainly of bluish grey to dark grey-brown, silty clay which weathers brown. Subordinate beds and laminae of silt and fine-grained sand are relatively common at the base and top of the formation, and there may be impersistent beds of well-rounded, black, flint pebbles up to 10 cm thick, and sporadic flint pebbles. Beds of calcareous cementstone concretions up to 0.4 m in diameter are present, some of them containing septarian calcite (calcium carbonate) veins. Phosphate nodules occur rarely. Glauconite, in the form of small pellets and microcrystalline grains, is quite common in some of the more sandy beds and locally at other horizons. Pyrite is disseminated throughout the rock as a replacement of fossil shell debris and as nodules up to 30 mm in diameter.

The weathered zone, in which the clay is oxidised to brown hues, varies from less than 1 m thick beneath river terrace gravels to around 5 m at outcrop. In a zone up to about 8 m below this, the reaction of acidic groundwater on calcareous fossil debris produces selenite (calcium sulphate) crystals that are both finely disseminated and within fissures within the clay.

The detailed succession of the London Clay in this district is not known, but is likely to be similar to that in the adjacent Windsor and Bracknell district to the north (King, 1981; Ellison and Williamson, 1999). Five sedimentary cycles, denoted as units A to E in ascending sequence, have been recog-

nised in the London Clay by King (1981). An idealised cycle commences with a thin bed of clay containing coarse sand grains, scattered well-rounded, black, flint pebbles and glauconite grains. Clays that, by gradation, become increasingly silty and culminate in fine sand overlie it. In regional terms, the sand content of each cycle increases towards the west. Numerous seams of calcareous nodules up to 100 mm in diameter occur within the lowest 20 m of London Clay. Elsewhere, correlation has utilised thin beds containing rounded black flint pebbles laid down during widespread transgressions at the base of King's units B, D and E.

The top part of the London Clay (corresponding to the top part of unit E), is mapped as the **Claygate Member** (ClgB) in the east of the district where it has also been interpreted in borehole records. It gives rise to relatively loamy soil and in places forms a small positive feature associated with a weak spring line. In terms of lithology the Claygate Member is transitional between the clay-dominated London Clay and the succeeding sand of the Bagshot Formation. It comprises a variable sequence of thinly interbedded, finely laminated, fine sands and clayey silts and clayey fine sands. At its base is a sand unit generally about 0.3 to 0.5 m thick. Where positively identified in the east of the district the total thickness is up to 6 m. In the west of the district, the top part of the London Clay is rather sandy and of a similar lithology to the Claygate Member, but it proved impractical to map consistently.

Thickness: decreases towards the west and south of the district. In the area around Wisley Common and in the more northern and central areas around Bisley and Pirbright, some 116 m are present. The thickest London Clay is 130 m, in the Woking area. In the west of the district, around the Hog Hatch and Upper Hale areas, to the north of Farnham, 90 to 110 m are proved (see also Lyons, 1887), the top 10 to 15 m of which locally resembles the facies of the Claygate Member.

BRACKLESHAM GROUP (BrB)

As a result of recently drilled cored boreholes, the Bracklesham Group succession is much better known than hitherto and the descriptions given below are therefore more detailed than those of other formations in this account.

Bagshot Formation (BgB)

This formation is equivalent to the Lower Bagshot Sands of Prestwich (1847) and the Bagshot Beds of Dewey and Bromehead (1915) and Dines and Edmunds (1929). It is correlated with the Wittering Formation in the Hampshire Basin (Edwards and Freshney, 1987). The top part of the formation is the **Swinley Clay Member** (SwC) (Ellison and Williamson, 1999). This overlies a sequence, dominantly of sand, that forms the bulk of the Bagshot Formation (undifferentiated).

Distribution: crops out extensively north of a line from Heath End, north of Farnham, to Cobham, the largest spread being in the Woking area. Characteristically, it forms rolling land with low rounded hills and small escarpments used largely for arable farming but with some areas of uncultivated sandy heathland. The Swinley Clay Member crops out from Normandy [SU 938 527] westwards to near Hale [SU 826 490] on the lower escarpment slopes of the Windlesham Formation.

Lithology: pale yellow-brown and grey, cross-bedded, fine- to medium-grained, quartz sands occur, locally with interbedded lenticular clay-dominated beds (formerly described as 'pipeclay'). Sections in the Aldershot area were described by Irving (1885), Monckton and Herries (1886), Lyons (1887) and Dines and Edmunds (1929).

There are a number of minor exposures of yellow and ochreous brown, iron-stained, cross-bedded, fine-grained, sands on the slopes north-east of Caesar's Camp west of Aldershot. The foresets dip consistently northwards.

The **Swinley Clay Member** is commonly grey to lilac-grey clay and silt. The base, on the underlying sands of the formation is sharp but irregular; the top is marked by a burrowed surface. The Mytchett No. 2 Borehole (Figure 6) proved laminated clays and silts with some burrowing and extensive postdepositional bleaching which decreases in intensity with depth. There is a prominent bed of lignite from 61.07 to 61.17 m depth and scattered rhizolith horizons indicating in-situ plant growth. Below about 61.80 m there is less regular lamination, and the clays are more massive with higher silt and sand content.

Strata now known as Swinley Clay are recorded in sections and illustrated by Prestwich (1847), Irving (1885), Monckton and Herries (1886), Lyons (1887) and Dines and Edmunds (1929). The most important of these were railway cuttings in Aldershot that are now degraded and overgrown [SU 872 507]. A section previously recorded by Dines and Edmunds (1929, p.96) at Sunny Hill [SU 8479 5059] currently exposes the basal 0.5 m of Swinley Clay and the top metre of sandy Bagshot Formation.

Thickness: about 45 m in the Normandy Borehole; at least 36 m in the Woking area, thinning in an easterly direction to only 10 to 15 m around Worplesdon; 15 to 20 m around Farnborough; 19 to 25 m near Aldershot. The Swinley Clay Member is 3.63 m thick in the Mytchett No. 2 Borehole (Figure 6) but mapping suggests up to 8 m at outcrop.

Windlesham Formation (Wi)

The Windlesham Formation (Ellison and Williamson, 1999) comprises all but the lowermost part of the former Bracklesham Beds of Dewey and Bromehead (1915) and Dines and Edmunds (1929), and the Middle Bagshot Sand of Prestwich (1847).

Distribution: in the west of the district the formation forms much of the low ground between Aldershot and Fleet. It also forms low escarpment and dip slope features within Aldershot. East of the Blackwater valley it forms large spreads of ground between Pirbright and Bisley, also occupying higher ground, for example at Cobbett Hill and Worplesdon.

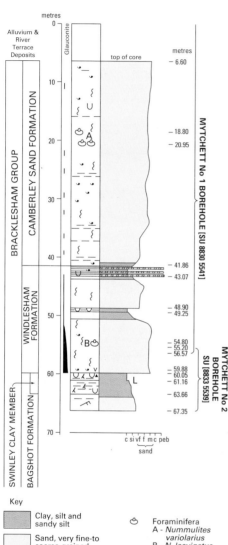

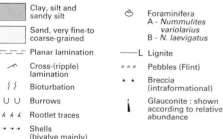

Lithology: the basal burrowed contact penetrates at least 0.7 m into the underlying Swinley Clay. The principal lithology is dark green to blue-green, bioturbated, fine- to medium-grained sand with subordinate paler greenish brown sands and clay laminae. Some of the beds comprise up to 70 per cent coarse sand-grade glauconite pellets. The sand becomes finer grained and generally less glauconitic and more clayey upwards. Locally at the top of the formation, there are thin beds of organic clay and flint pebbles, the latter probably equivalent to the Stanners Hill Pebble Bed of the Windsor and Bracknell district (Ellison and Williamson, 1999, p.13). The Mytchett boreholes (Figure 6) provide a detailed section of the formation. Unusually, because of the high water table in the Blackwater valley, a calcareous fauna is preserved. Of particular interest is the occurrence, at a depth of about 54.8 m, of *Nummulites laevigatus.* This large foraminiferid flourished during the warmest climatic interval of the Cenozoic (Purton and Brasier, 1999) and its presence suggests a direct correlation with the Earnley Formation of the Hampshire Basin sequence (Edwards and Freshney, 1987).

Former exposures were described by Prestwich (1847) and Dines and Edmunds (1929). The best exposures are now in the west of the district, where up to 5 m of variably glauconitic, clayey, medium- to coarse-grained sand are seen on Jubilee Hill north of Caesar's Camp [SU 8410 5099] and in gullies in the hilly area to the south [SU 8390 5068; SU 8403 5027]. East of Aldershot [SU 9292 5245] 2 m are exposed in a deep ditch section.

Thickness: highly variable, generally 6 to 25 m.

Camberley Sand Formation (CaS)

The Camberley Sand Formation (CaS) was defined by Ellison and Williamson (1999) to replace the term 'Barton Beds' as used by

Figure 6   Graphic sections of the BGS Mytchett No.1 and No.2 boreholes.

Dines and Edmunds (1929) and the Upper Bagshot Sands of Prestwich (1847).

Distribution: occupies substantial areas in the north-east of the district, with steep-sided outliers capped by river terrace deposits and head gravels. Boreholes indicate that the break of slope formerly taken as the base of the formation, particularly around Cove, Farnborough and Frimley, is caused by springs above clayey beds and/or beds of ironstone, several metres above the base.

Lithology: uniform, pale yellow-brown, sparsely to moderately glauconitic, bioturbated, fine-grained sands. There are sporadic thin beds of grey clay ('pipeclay') and pebbles are locally recorded from the lower beds. Exposures may show cross-bedding (foresets dip to the south-south-east at Frimley [SU 8948 5859]) and subhorizontal laminae and burrows as at Hawley Common [SU 8425 5837; SU 8424 5854; SU 8471 5848]. Red-brown staining and iron pan is mainly concentrated in the top metre below ground surface but may also extend down irregularly over several metres. The Mytchett No. 1 Borehole provides a reference section for the lowest 35 m of the formation (Figure 6). It consists of mottled greenish grey to olive-green and brown, moderately to well-sorted, fine- to very fine-grained quartz sand with scattered glauconite grains of medium sand grade. The quartz grains are mostly subrounded. Clayey streaks and lenses occur throughout and there are thin beds of sandy clays with burrows at the base. The sediments are mainly homogeneous and intensively bioturbated; faint lamination occurs in places and there are poorly developed upwards-fining units. Scattered bivalves, including *Lentipecten corneum*, and *Nummulites variolarius* indicate correlation (C King, personal communication) with the Selsey Formation of the Hampshire Basin (Edwards and Freshney, 1987)).

Thickness: up to 55 m; the top of the formation is not preserved.

Quaternary (Drift)

Quaternary deposits are mapped over about 15 per cent of the district. They include deposits of periglacial mass movement, river terraces and postglacial alluvium of the present-day drainage system.

The main outcrops were delineated during the primary survey at the 1:63 360 scale in the 1880s. These have been revised from field evidence such as open sections, topographical features, auger holes and borehole data.

The oldest Quaternary deposits are the **Caesar's Camp Gravel** that caps an area of small hills and plateaux, 3 km north of Farnham at about 175 to 187 m above OD, the highest ground in the district. The deposits, which rest with sharp contact on Palaeogene sands, are generally accepted as being fluvial and periglacial in origin, laid down by streams draining towards the north-east from chalk uplands (Clarke and Fisher, 1983). The base is obscured in most places by a veneer of solifluction deposits, but in exposures appears to be highly irregular. The maximum thickness proved is 4.8 m in a borehole in the centre of the outcrop [SU 8313 4935]. The base is in general at 175 to 184 m above OD and inclined slightly to the east. The deposit consists of unbedded and massively bedded cobble gravel with interbedded coarse sands. The clasts are dominantly of nodular flint, many of them larger than 0.2 m in diameter. The sand beds are mainly of subangular quartz with some flint. They include horizontal planar-bedded units, cross-bedded sand lenses and channel-fill sands and silts. The top part of the deposit contains involutions caused by postdepositional periglacial activity. Incorporated in these involutions are orange, brown and bright red-brown sands thought to have developed as part of a palaeosol profile (Clarke and Fisher, 1983). Locally, a bed of clayey silt 0.8 m thick, presumed to be aeolian in origin, disconformably overlies the involuted bed.

The **Netley Heath Beds** (N) are stratigraphically complex. The prevailing view on their origin is that they were laid down in

Pliocene time (about 1.8 Ma) in a marine transgression which reworked Tertiary (Palaeogene) sediments. Subsequently they have been cryoturbated, mixed with locally derived solifluction deposits and suffered collapse into chalk dissolution cavities. There is no preferential arrangement of these different lithologies, and stratification is generally absent because of intense mixing by cryoturbation. John (1980) re-evaluated the deposits on the basis of soil type at outcrop but the resulting lithological subdivisions proved impractical to map out. The Netley Heath Beds occur on the crest of the Hog's Back and on the Chalk dip slope, at elevations from 150 m above OD in the west of the outcrop [TQ 027 492] to about 223 m above OD in the east at Hackhurst Downs [TQ 100 490]. Old descriptions, based on sections in former pits (Dines and Edmunds, 1929), indicate two main lithologies. In the west, around Newlands [TQ 043 492], the deposit is dominantly a clayey gravel. The clasts are mainly of battered nodular flints up to small boulder (0.3 m) size, well-rounded black 'Tertiary' flints, a variable proportion up to 20 per cent coarse pebbles of Lower Greensand, quartz pebbles and boulders of sarsen. In the east of the district, around Mountain Wood [TQ 090 505] and Netley Heath [TQ 086 494], the deposit consists dominantly of sand with clasts similar to those in the west. Blocks of iron-cemented, pebbly sandstone (presumably reworked) containing a fauna similar to the Red Crag (Dines and Edmunds, 1929, p.112) are recorded from the base of the Netley Heath Beds in pits [at TQ 0837 4990]. The thickness of the deposits is generally in the order of 5 m, but may be 10 m in deep dissolution features in the underlying Chalk.

The principal outcrop of **Clay-with-flints** is on the escarpment and dip slope of the Upper Chalk in the east of the district north of Hackhurst Downs [TQ 10 49]; smaller outcrops are present on the eastern part of the Hog's Back ridge. On Pewley Down [TQ 028 492] they overlie the Netley Heath Beds. The deposit comprises stiff, brown to reddish brown clays and silty clays containing numerous nodular and well-rounded flint pebbles and cobbles. It is primarily a *remanié* deposit that is thought to have formed through a combination of collapse and modification of the original Palaeogene and Neogene cover and dissolution of the underlying Chalk.

The deposits of an ancient braided river draining in a northerly direction from the Hampshire Downs and the western part of the Weald are mapped as **River Terrace Deposits (Ninth Terrace to Seventh Terrace)**. This river formerly joined the ancestral River Thames that more than 0.5 million years ago flowed well to the north of its present course. Spreads of these terrace deposits occur on broad plateaux in the west of the region, mainly around Yateley Common [SU 825 588], Chobham Ridges [SU 908 596] and Romping Down [SU 910 537]. The sediments consist mainly of flints derived from chalk downland, a smaller number of Lower Greensand chert and sandstone clasts and Tertiary rounded flint pebbles (Figure 7; Gibbard, 1982).

The oldest river terrace deposits, the **Surrey Hill Gravel (Ninth terrace)**, occur at about 120 to 124 m above OD. About 10 m and 18 m below are the **Eighth and Seventh River Terrace Deposits**, respectively. The Eighth Terrace Deposits consist of gravels with thick sand-dominated units locally. The Seventh Terrace Deposits are broadly similar.

After the main Anglian glaciation, the River Thames was diverted to near its present course. The modern Blackwater and Wey tributary valleys were established at this time, and aggradation of a new suite of deposits, the **post-diversionary River Terrace Deposits**, continued until Holocene time about 10 000 years ago. The dissected remnants of these deposits occur generally up to about 20 m above the present-day floodplains. They are preserved in a 'staircase' caused by the progressive down-cutting of the river channels that is probably driven by continuing gentle uplift of the region.

The terrace deposits of the River Blackwater (Figure 7) are numbered consistent with terraces recognised farther downstream in the adjacent Reading district

(Mathers, 2000) and Windsor and Bracknell district (Ellison and Williamson, 1999). Correlation of the oldest (highest) terrace deposits is however uncertain and they are denoted **River Terrace Deposits undifferentiated**. **River Terrace Deposits 4** forms a sizeable spread about 8 m above the valley floor at Farnborough. **River Terrace Deposits 2/3** embrace the height range of two terraces downstream from this district. The composition of these river terrace deposits downstream from Badshot Lea [SU 872 484] is shown on Figure 7. Those south-east of Badshot Lea, in the vicinity of Farnham, were described in detail by Dines and Edmunds (1929), Worssam (1973) and Bryant et al. (1983). They consist of highly ferruginous sandy gravel, with a clay matrix in the upper 3 m.

The upstream part of the former Blackwater valley is now drained by the River Wey as a result of 'river capture' between Farnham and Badshot Lea following deposition of River Terrace Deposits 2/3, which are possibly of Devensian age (see Worssam, 1973; Bryant et al., 1983). One possible cause of the diversion may have been the development of icings (frozen spring water) forming a dam at the boundary between the Gault and Chalk during a cold phase when the water table was elevated (see a review in Mathers, 2000).

River Terrace Deposits named **Lynch Hill Gravel, Taplow Gravel and Kempton**

Figure 7   Classification and composition of River Terrace Deposits (after Clarke and Dixon, 1983).

River system	River Terrace deposits	Equivalent names and former names	Clast type and percentage					
			Angular flint	Rounded flint	Lower Greensand	Quartz	Others	Thickness (m)
	Caesar's Camp Gravel		90 to 91	7 to 9	0 to 1	0 to 1	0 to 3 (ironstone)	up to 5
Tributary of the pre-diversionary River Thames	Surrey Hill Gravel (Ninth)	Plateau Gravel * Easthampstead Gravel † 10th and 11th terraces ‡						2 to 4
	Eighth	Plateau Gravel * Fox Hills Gravel † 8th terrace ‡	70 to 75	17 to 20	8 to 9	<1		up to 5
	Seventh	Plateau Gravel *	82	6	11	<1		<2 to 5
River Blackwater		Warren Heath Gravel †						
	Undifferentiated	5th terrace ‡ 2nd terrace *						2 to 4
	Fourth	4th terrace ‡ 2nd terrace *	83 to 88	3 to 5	4 to 11	2 to 2.4	0 to 2.5 (chalk)	up to 5
	Second to Third	2nd terrace and 3rd terrace ‡ 1st terrace * Wrecclesham Gravel †						3 to 6
River Wey (north)	Lynch Hill Gravel	2nd terrace *						generally <4
	Taplow Gravel	1st terrace *						
	Kempton Park Gravel	1st terrace * Byfleet Gravel †						
River Wey(south)	Undifferentiated First	1st and 2nd terrace						

* Dines and Edmunds, 1929;   † Gibbard, 1982;
‡ Clarke et al., 1979;   Clarke and Dixon, 1981;   + Bryant et al., 1983

Park Gravel occur in the Wey valley north of Guildford, up to 30 m above present river level. They are correlated with the River Thames terrace sequence to the north (Gibbard, 1985; Ellison and Williamson, 1999) and are mainly gravelly sand (Dines and Edmunds, 1929).

The River Wey terrace deposits south of the Hog's Back cannot be correlated with certainty with those to the north. Two terrace deposits are recognised, up to 15 m above river level. The higher one, **River Terrace Deposits undifferentiated**, being typically separated from the lower, **First River Terrace Deposits**, by a rock step, for example at Elstead. Both terrace deposits contain a high proportion of sand, particularly in the large outcrops around Peasmarsh [SU 995 462] where mammal bones have been found and locally the lowest bed is peat (Dines and Edmunds, 1929, p.138). The gravel component consists of ironstone fragments (carstone) derived from the Folkestone Formation, Lower Greensand cherty sandstone and angular flint.

The **Alluvium** deposits were laid down on the present floodplains in the last 8000 years. They are generally 2 to 3 m thick, but locally up to 4 m in the River Wey valley. They consist principally of soft, unconsolidated, grey and brown sand, silt and clay, locally with peat beds recorded up to 1.83 m thick in boreholes. Underlying the alluvium of the large rivers are sands and gravels. Deposits up to 3 m thick were proved in boreholes for the M25 motorway in the lower reaches of the River Wey valley [TQ 063 599] (Gibbard, 1985, p.70). A bed of peat in the top of the sub-alluvial gravel in the Blackwater valley 4 km north-west of this district, is dated as 7891 years BP (Clarke et al., 1979).

Periglacial activity in a succession of cold periods and more recent downwash was responsible for the formation of **Head** and **Head Gravel** deposits. Head is widespread in the district but has been mapped mainly in valleys where locally more than 2 m may occur beneath concave slopes.

The lithology of head is determined largely by the parent materials upslope.

Head gravel commonly occurs at higher elevations than other head. The outcrops are the dissected remnants of solifluction lobes and periglacial apron fans that were widespread during erosion of extensive spreads of older gravels and Clay-with-flints. The lithology, generally clayey gravel, is dependent on the Quaternary source material and, to a lesser extent, the local bedrock. The deposits are mainly less than 2 m thick. Head gravel occurs in four main situations:

- isolated hilltops between Farnham [SU 82 49] and Pirbright [SU 94 55], largely overlying the easily eroded Camberley Sand; lithology comprises mainly sand with flint gravel
- shallow dry valleys that drain down the Chalk dip slope on to the London Clay, mainly between Burpham and Effingham Junction [TQ 08 56; TQ 045 528]; lithology comprises clay and sand with angular and well-rounded flint
- linear outcrops in the Wrecclesham area [SU 840 453] south of Farnham where the gradient of the base of the deposits is steeper than that of the base of the adjacent river terrace deposits; lithology comprises clayey gravel, mainly flint
- in the River Wey valley about 30 to 35 m above river level; lithology comprises flint gravel in clayey sand matrix

Landslips are likely to develop on slopes of 7° or greater, underlain by clay, in particular the Weald Clay, Atherfield Clay, Gault and London Clay. They may occur also on gentler slopes where the ground is saturated by springs and has high pore pressures and consequent low strength. Such areas may also have been subjected to landslip during periglacial episodes but are now relatively dry and stable. The landslips that have been mapped with certainty in the district occur in London Clay on valley sides that have been oversteepened during periods of high surface runoff from the Chalk [SU 8165 4755; SU 8270 4767; SU 9135 5042; SU 985 502].

Artificial deposits and worked ground

Made ground is shown in areas where material is known to have been deposited by man upon the natural ground surface. The main categories are spoil from mineral extraction, building and demolition rubble and waste in raised landfill sites. Other extensive areas of made ground include road and railway embankments; for clarity many such linear areas have been omitted from the 1:50 000 scale map.

Worked ground is where natural materials are known to have been removed, for example in quarries and pits, and excavation for roads and railways. In this district some of the large former gravel pits in the River Blackwater valley are not backfilled and shown as worked ground.

Infilled ground is shown where the natural ground has been removed and the void partly or wholly backfilled with man-made deposits. In this district most of these are pits in the Folkestone Formation south of Tongham [SU 87 47]; others are near Clandon Park [TQ 039 507] in the Upper Chalk, and at Albury [TQ 056 483] in the Folkestone Formation. Where old quarries and pits have been wholly or partly filled there may be no unequivocal surface indication of the extent of the backfilled area. In such cases the boundaries of these sites are taken from archival sources and may be imprecise.

Disturbed ground relates to ill-defined small mineral workings and spoil heaps that have not been individually mapped.

Landscaped ground is where there has been minor cut and fill (for example, playing fields) or general earth moving such as on golf courses and large-scale industrial and business parks.

Structure

Structures in the concealed strata have been identified largely from seismic reflection profiles and are inferred from gravity data, as illustrated on the 1:50 000 scale geological map. The district is dominated by east–west trending structures related to the faulted southern margin of the London Platform and the northern edge of the Weald Basin. Throw on the principal faults in the Jurassic and Cretaceous strata is mainly down to the south (see inset figures on the geological map) with the result that 'basement' lies at progressively greater depths southwards.

The main surface structure in the district is the north-facing Hog's Back Monocline. It formed due to compressive stress causing reverse movement on the principal faults at depth along the margin of the London Platform. The representation of this structure was described by Smalley and Westbrook (1982) and is illustrated on the cross-sections of the 1:50 000 scale map (after Lake and Shephard-Thorn, 1985). At outcrop, the fold affects mainly the Gault, Upper Greensand and Chalk, and to a lesser extent the Folkestone Formation. The strata involved locally dip up to 55° to the south, the steepest values being in the Chalk. Strike faults with throws in excess of 75 m occur at several locations along the Hog's Back. In particular the narrowness of the Gault outcrop east of Farnham is thought to be due to attenuation caused by internal shearing parallel to the main Hog's Back structure (see Lake and Shephard-Thorn, 1985 for details). The regional structure was also described by Ruffell (1992a).

North of the Hog's Back, the Chalk and Tertiary strata are gently folded along roughly east–west axes into a broad synclinorium that is part of the larger London Basin structure. Dip values are generally less than 2°. This is illustrated by the inset map of contours on the top of the Chalk Group in the margin of the geological map.

South of the Hog's Back, the Lower Cretaceous strata are also gently folded in a gentle east–west anticlinal fold axis through Peasmarsh. Dines and Edmunds (1929, p.47) also identified small folds in the Guildford–Tyting Farm [TQ 02 49] and West Horsley [TQ 08 52] areas.

3 Applied Geology

Geological factors may be important in influencing the suitability, siting and nature of future development. By giving consideration to information about the ground at an early stage in the planning process it may be possible to mitigate some of the problems commonly encountered. In addition, knowledge of the ground enables better decisions to be made concerning conservation of natural habitats, modelling of aquifers and issues of contamination.

Figure 8 lists the types of mineral workings in the district, and Figure 9 identifies the main issues regarding ground conditions relating to each formation.

Surface mineral workings

These include former or active pits and quarries that have not been backfilled. The largest have been excavated in river terrace deposits and the Folkestone Formation. They are an important resource that may be suitable for waste disposal, for re-opening as a source of minerals, or may be redeveloped as sites of educational, recreational or wildlife conservation value.

Water resources

Information about historical well-yields in the district is given in Flatt et al. (1976). The Chalk is the most important aquifer. Despite its high porosity, boreholes yield significant amounts of water only when they encounter fissures. The degree and distribution of fissure permeability therefore determines the magnitude of yield. The aquifer in this district is only of limited use because the narrow outcrop provides relatively low recharge, poor flushing potential and thus poor water quality. There is only a narrow tract about 5 km wide north of the Hog's Back, including Aldershot and Guildford, where water can be efficiently extracted. To

the north of this, the Chalk is deeper than 200 m below the surface (see inset map included on the 1:50 000 scale geological sheet) and fracture permeability is poor.

The Lower Greensand (Folkestone, Sandgate, Bargate and Hythe formations) is an important aquifer but with lower yields than the Chalk. Permeability is greatest in the Folkestone Formation. The aquifer as a whole is vulnerable to surface pollution, in particular in the area south-east of Farnham where the Folkestone and Sandgate formations are unconfined. In this area there is a high recharge potential due its wide outcrop and few major drainage courses. An additional factor to take into account is the presence of water of relatively low pH, undersaturated with respect to carbonates, above the saturated zone. At Crooksbury Common [SU 8878 4510], on the Folkestone Formation crop, Edmunds et al. (1992) showed that some metallic elements which dissolved from the sand matrix in the surface layers may persist in a relatively high concentration in the top 15 m of the saturated zone. This metal concentration is attenuated during percolation through the unsaturated zone. The Folkestone Formation aquifer may in the future be considered for use in storage and recovery projects.

Foundation conditions

There are few problems associated with ground stability in the district. In valleys, alluvium, head and peat can pose problems because of lithological variability and generally low bearing capacity. Peat occurs locally in the alluvial sequences in the Blackwater and Wey valleys.

The locally unpredictable nature of the Palaeogene strata may present problems for ground investigations and excavations (Figure 9). Loose sands prone to erosion and gullying are characteristic of the

Camberley Sand, Bagshot and Folkestone formations. The development of springs at the base of these formations, especially after periods of heavy rain, may cause erosion and burst-out of the sands.

All clay-dominated formations are susceptible to significant volume change, **ground heave**, as a result of changes in moisture content. The Gault and London Clay are most affected, the Weald Clay slightly less so. The seasonal rainfall, exacerbated by vegetation, particularly trees, is the most important control on shrinking and swelling (Driscoll, 1983). The mechanism involves absorption of water, mainly by the clay mineral smectite, during wet periods, and loss of water during dry seasons causing cracking and desiccation of the

Figure 8   Summary information on the mineral resources of the Guildford district.

Mineral resource	Geological unit	Principal use	Activity
Sand and gravel	River Terrace Deposits	aggregate	Extensive former workings in Wey and Blackwater valleys; most recent workings are near Ripley [TQ 02 56] and near Broadwater Farm [SU 984 453]
Sand and gravel	Pre-diversionary Thames River Terrace Deposits	aggregate	Small pits [e.g. SU 907 607; 905 517]
Sand and gravel	Head gravel	aggregate	Small former pits as at Pondtail [SU 82 54]; south of Bastion Hill [SU 9160 5275]
Sand and gravel	Caesar's Camp Gravel	aggregate	Disused pits at Ewshot [SU 825 494]
Sandstone	Bargate Formation	building stone aggregate	Extensive former quarrying in valleys near Godalming: [SU 9605 4255]; Crown Pits [SU 976 432]; Shackstead Lane [SU 9655 4332]
Clay	Windlesham Formation	brickmaking pottery	Knaphill [SU 971 586]; Normandy [SU 9100 5135]
Clay	London Clay	brickmaking	Upper Old Park Brickworks [SU 831 481]; Aldershot Brick and Tile works [SU 8665 4955]; Wanborough [SU 9375 5050]; Ash [SU 8845 5000; 8665 5960]; north of Guildford [SU 976 522; 9773 5178; 9770 5196; 9783 5200; 9824 5190]
Clay	Reading Formation	brickmaking pottery	Guildford Park Brickworks [SU 986 498]
Clay	Gault Formation	brickmaking pottery	Around Wrecclesham [SU 8265 4480; 833 440]
Clay	Weald Clay and Atherfield Clay	brickmaking	Littleton [SU 983 469]; Peasmarsh [SU 9913 4660]; north of Wonersh [TQ 0145 4630]
Chalk	Chalk Group	agricultural lime	Numerous old pits, the largest at Clandon [TQ 040 507] and on the Hog's Back [SU 946 483; 952 483]
Brickearth	River Terrace Deposits	brickmaking	Former pits at Farnham [SU 832 463]
Ironstone	Folkestone Formation	walling and roofing aggregate	No specific working identified
Sand	Folkestone Formation	building sand and for use in tarmac	Currently active pits at Albury [TQ 057 483] and Wrecclesham [SU 825 452]; Numerous disused pits, principally around Seale [SU 873 473 to 898 475]
Sand	Hythe Formation	building sand	Old pit south of Guildford [SU 9890 4775]; active pit near Pitch Hill [TQ 0795 4253]

surface clay layers. The seasonal movement of surface clay may be in the order of several centimetres, and on sloping ground this can lead to significant creep.

High concentrations of **sulphate** in groundwater can weaken concrete foundations that are not designed to resist the chemical attack. In the weathered zone near the surface, pyrite is oxidised to give sulphate ions in solution. In the clay-dominated formations, particularly London Clay and Gault (Forster et al., 1995), the calcium carbonate present may react with the sulphate to precipitate selenite crystals. This involves an eightfold increase in volume compared with the original pyrite and can cause disruption and weakening of the strata. Furthermore, if the selenite is subjected to weathering, sulphuric acid is produced, which reacts with cement not designed to be sulphate resistant, causing it to break down. Selenite can occur in grey (apparently unweathered) clays at depths of up to 10 m below the surface.

Mass movement and slope stability

Few landslips are mapped in the district (see p.20). In adjacent districts, the Lower Greensand escarpment, formed by cemented beds in the Hythe Formation, is extensively landslipped with much of the failure at the contact with the underlying Atherfield and Weald clays. In this district there is only one minor landslip on the escarpment, involving head deposits overlying the Hythe Formation south of Wonersh [TQ 014 427], although it is likely, that considerable mass movement has occurred as a result of periglacial activity.

Many slopes greater than 3° in the clay-dominated formations have a veneer of head deposits that may not be shown on the geological maps. Culshaw and Crummy (1991) suggested that on London Clay particularly, these too should be considered as potentially unstable. The head, composed of redeposited clay may contain relict shear surfaces and is likely to have shear strengths at, or close to, the residual value. Reactivation of the shear surfaces may occur if the slopes are undercut, loaded or saturated.

Chalk dissolution

Swallow holes or sinkholes are a feature of the Chalk in the Chiltern Hills and North Downs. They generally occur in valleys and around the margins of Tertiary outliers, locations that are most liable to the development of karst features in the chalk. They are not known in the major river valleys in this district but small hollows occur at Farnham Park [SU 8460 4798; SU 8477 4796] and around the periphery of the Tertiary outlier at Hook Wood [TQ 076 509; TQ 080 508]. Beneath the superficial deposits, along the crest of the Chalk escarpment east of Guildford, the chalk surface is highly irregular, with pipes and hollows filled with intermixed pebbly clays and sands from the Clay-with-flints and Netley Heath Beds. Evidence elsewhere in the North Downs suggests that some pipes may be more than 10 m deep.

Flooding

The floodplains adjacent to streams and rivers are shown as alluvium on geological maps. They are, as their name suggests, liable to flood following periods of exceptional rainfall. To reduce the frequency of flooding, the Wey and Blackwater rivers have been canalised in places. Even so, the Wey in particular regularly overtops its banks on to the adjacent flood meadows. During high floods, low-lying developed areas of Guildford, and parts of Woking that are underlain by river terrace deposits, are also at risk. The Wey flows in part over Lower Greensand that in times of flood is easily eroded by the river. Subsequent deposition of this sediment changes the river flow characteristics and ecology of the riverbed. To alleviate this, the sand is trapped and periodically removed from several monitoring points along the river.

The alluvium along the Cove Brook tributary of the River Blackwater [85 56] has been largely built over, and a floodwater storage scheme reduces the risk of flooding.

Figure 9   Summary information on ground constraints

ARTIFICIAL DEPOSITS

Geological unit	Considerations
Worked ground	variable foundation conditions unstable sides of old workings
Made ground Infilled ground	variable foundation conditions leachate and methane production from waste
Landscaped ground Disturbed ground	slope instability variable foundation conditions

QUATERNARY DEPOSITS

Geological unit	Considerations
Landslip	unstable land
Head & Head gravel	variable lithology; perched water table; clay-dominated deposits have low strength
Alluvium	may contain peat flooding
River Terrace Deposits	high water table
Netley Heath Beds	highly irregular base
Clay-with-flints	large flints in clay matrix; perched water table in sandy clays; highly irregular base

SOLID GEOLOGY

Geological unit	Considerations
Camberley Sand Formation	unstable in excavations with high water table locally caused by iron pan development
Windlesham Formation	local perched water table variable lithology high glauconite content in places
Bagshot Formation	local perched water table
London Clay Formation	ground heave sulphate-rich groundwater landslip
Lambeth Group (Reading Formation)	variable lithology
Thanet Sand Formation	basal bed contains large nodular flints
Chalk Group	groundwater protection required possibility of undocumented former pits dissolution cavities and sinkholes
Gault Formation	ground heave ancient landslips on low-angle slopes
Folkestone Formation	ironstone (carstone) concretions and veins acidic surface groundwater
Sandgate Formation	ferruginous concretions acidic surface groundwater local perched water table caused by beds of ironstone
Bargate Formation	calcareous cemented sandstone beds chert layers
Hythe Formation	hard cherty sandstone beds at top
Atherfield Clay Formation	ground heave
Weald Clay Formation	ground heave

Information sources

Further geological information held by the British Geological Survey relevant to the Guildford district is listed below. Enquiries concerning geological data for the district should be a addressed to the Manager, National Geological Records Centre, BGS, Keyworth. Geological advice for this area should be sought from the Geological Enquiry Service, BGS, Keyworth.

Information on BGS products is listed in the current *Catalogue of geological maps and books* and on BGS website (addresses on back cover).

Books

British Regional Geology
London and the Thames valley, 4th edition, 1996
The Wealden district, 4th edition, 1965
The Hampshire basin and adjoining areas, 4th Edition, 1982

Memoirs
Aldershot and Guildford (Sheet 285), 1929*
Windsor and Chertsey (Sheet 269), 1915*
Reigate and Dorking (Sheet 286), 1933*
Haslemere (Sheet 301), 1968*
Basingstoke (Sheet 284), 1909*
* denotes out of print

Sheet Explanation
Geology of the Windsor and Bracknell district (Sheet 269), 1999

Water Supply, Well Inventory Series
Records of wells in the area around Aldershot: Inventory for geological Sheet 285.

Maps

Geology

1:1 500 000
Tectonic map of Britain, Ireland and adjacent areas, 1996

1:1 000 000
Pre-Permian geology of the United Kingdom, 1985

1:625 000
United Kingdom (South Sheet) Solid geology, 1979; Quaternary geology, 1977

1:250 000
51N 02W Chilterns, Solid geology, 1991

1:50 000
Solid and Drift
  Sheet 268 Reading
  Sheet 269 Windsor, 1999
  Sheet 270 South London
  Sheet 284 Basingstoke, 1980
  Sheet 285 Guildford, 2001
  Sheet 286 Reigate, 1978
  Sheet 300 Alvesford
  Sheet 301 Haslemere, 1981

1:10 000
Details of the original geological surveys are listed on editions of the 1:50 000 geological sheets. Copies of the fair-drawn maps of the earlier surveys may be consulted at the BGS Library, Keyworth.

The maps covering the 1:50 000 Series Sheet 285 Guildford are listed below together with the surveyors' initials and the date of survey. The surveyors were C R Bristow, R A Ellison, A J Humpage, R D Lake, E R Shephard-Thorn, A Smith, P J Strange, I T Williamson and J A Zalasiewicz.

The maps are not published but are available for public reference in the libraries of the British Geological Survey at Keyworth and Edinburgh and the BGS London Information Office in the Natural History Museum, South Kensington, London. Uncoloured dyeline sheets or photographic copies are available for purchase from the BGS Sales Desk.

Geophysical maps

1:1 500 000 & 1:250 000
See Sheet 285 Guildford for coverage.

1:50 000
A geophysical information map (GIM) at the scale of 1:50 000 is available for the

27

BGS digital databases, including Bouguer
gravity and aeromagnetic anomalies and
locations of data points, selected boreholes
and detailed geophysical surveys.

Geochemical maps
1:625 000
Methane, carbon dioxide and oil suscepti-
bility, Great Britain — South Sheet, 1995.
Radon potential based on solid geology,
Great Britain — South Sheet, 1995.
Distribution of areas with above national
average background concentrations of
potentially harmful elements (As, Cd, Cu,
Pb and Zn), Great Britain — South Sheet,
1995.

Hydrogeological maps
1:625 000
Hydrogeological map of England and
Wales, 1977
1:100 000
Hydrogeological map of the Chalk and
Lower Greensand of Kent (including part
of hydrometric area 40), 1970
  Sheet 1: Chalk, Regional hydrological
  characteristics and explanatory notes.
  Sheet 2: Folkestone Beds and Hythe Beds
Groundwater vulnerability of west
London, Sheet 39, 1994
Groundwater vulnerability of West Sussex
and Surrey, Sheet 45, 1995

Minerals maps
1:1 000 000
Industrial minerals resources map of
Britain, 1996

BGS reports
Additional reports not included in
References (p.29, 30). These and most of the
references listed are available for consulta-
tion at the Library, BGS, Keyworth.

LAKE, R D, and SHEPHARD-THORN, E R. 1982.
The surface geological structure of the
Hog's Back and adjoining areas. *Institute
of Geological Sciences Report for the
Department of Energy.*
WOODS, M A. 1998. Review of BGS macro-
fossil collections from the Chalk of the

Reigate Sheet (286): Part 1. *British Geological
Survey Technical Report*, WH/98/158R.

Documentary collections
Boreholes
Borehole data for the district are catalogued
in the BGS archives (National Geological
Records Centre) at BGS Keyworth on indi-
vidual 1:10 000 scale sheets. For further
information contact: The Manager, NGRC,
BGS, Keyworth.
BGS hydrogeology enquiry service; wells,
springs and water borehole records.
British Geological Survey, Hydrogeology
Group, Maclean Building, Crowmarsh
Gifford, Wallingford, Oxfordshire OXO 8BB.
Telephone 01491 838800. Fax 01491 692345.

BGS Lexicon of named rock unit defini-tions
Definitions of the named rock units shown
on BGS maps, including those on the
1:50 000 Series 285 Guildford Sheet are held
in the Lexicon database. This is available
on Web Site *http://www.bgs.ac.uk* Further
information on the database can be
obtained from the Lexicon Manager at BGS,
Keyworth.

BGS photographs
Copies of these photographs are deposited
for reference in the BGS libraries in
Keyworth and Edinburgh. Colour or black
and white prints and transparencies can be
supplied at a fixed tariff.

Materials collections
Palaeontological collections
Macrofossil and micropalaeontological
samples collected from the district are held
at BGS, Keyworth. Enquiries regarding all
the fossil material should be directed to the
Curator, Biostratigraphy Collections, BGS,
Keyworth.

Other relevant collections
Groundwater licensed abstractions, Catchment Management Plan and landfill sites
Information on licensed water abstraction
sites, for groundwater, springs and reser-

Map	Name	Author	Date
SU84NW	Farnham	RDL, ITW	1981, 1998
SU84NE	Weybourne and Tongham	RDL, ITW	1981, 1998
SU84SW	Rowledge and Frensham	CRB, ITW	1997, 1998
SU84SE	Tilford	ITW	1998
SU85NW	Cove	ITW	1997
SU85NE	Farnborough and Frimley	ITW	1997
SU85SW	Fleet	ITW	1997
SU85SE	Aldershot	ITW	1997
SU86SW	Crowthorne	PJS	1996
SU86SE	Camberley	PJS	1996
SU94NW	Puttenham	RDL, RAE	1981, 1998
SU94NE	Guildford	ERST, AJH	1981–82, 1998
SU94SW	Elstead	RAE	1998
SU94SE	Godalming	AJH	1998
SU95NW	Pirbright Common	RAE	1998
SU95NE	Knaphill	AS	1997
SU95SW	Normandy	RAE	1998
SU95SE	Stoughton	AS	1997
SU96SW	Bagshot	ITW	1996
SU96SE	Chobham	AS	1996
TQ04NW	Chilworth	ERST, AJH	1981–82, 1998
TQ04NE	Shere	ERST, AS	1982, 1998
TQ04SW	Wonersh and Bramley	AJH	1998
TQ04SE	Ewhurst	AS	1998
TQ05NW	Woking	AS	1997
TQ05NE	Ockham	AS	1998
TQ05SW	Burpham	AJH	1998
TQ05SE	West Horsley	AS	1998
TQ06SW	Ottershaw	RAE	1996
TQ06SE	Weybridge	RAE	1996
TQ14NW	Westcott	RDL	1982
TQ14SW	Holmbury St. Mary	JAZ	1981
TQ15NW	Stoke D'Abernon and Fetcham	ITW	1998
TQ15SW	Great Bookham	ITW	1999
TQ16SW	Esher and Cobham	ITW	1994

voirs, Catchment Management Plans with surface water quality maps, details of aquifer protection policy and extent of Washlands and licensed landfill sites are held by the Environment Agency.

Earth science conservation sites

Information on the Sites of Special Scientific Interest within the Guildford district is held by English Nature, Headquarters and Eastern Region, Northminster House, Peterborough.

References

BRISTOW, R, MORTIMORE, R, and WOOD, C. 1997. Lithostratigraphy for mapping the Chalk of southern England. *Proceedings of the Geologists' Association*, Vol. 108, 293–315.

BRYANT, I D, GIBBARD, P L, HOLYOAK, D J, SWITSUR, V R, and WINTLE, A G. 1983. Stratigraphy and palaeontology of Pleistocene cold-stage deposits at Alton Road Quarry, Farnham, Surrey, England. *Geological Magazine*, Vol. 120, 587–606.

CASEY, R. 1961. The stratigraphical palaeontology of the Lower Greensand. *Palaeontology*, Vol. 3, 487–622.

CLARKE, M R, and DIXON, A J. 1981. The Pleistocene braided river deposits in the Blackwater valley area of Berkshire and Hampshire, England. *Proceedings of the Geologists' Association*, Vol. 92, 139–157.

CLARKE, M R, and FISHER, P F. 1983. The Caesar's Camp Gravel — an early Pleistocene fluvial periglacial deposit in southern England. *Proceedings of the Geologists' Association*, Vol. 94, 345–355.

CLARKE, M R, DIXON, A J, and KUBALA, M. 1979. The sand and gravel resources of the Blackwater valley (Aldershot) area. Description of 1:25 000 sheets SU85, 86 and parts of SU84, 94, 95 and 96. *Mineral Assessment Report of the Institute of Geological Sciences*, No. 39.

DEWEY, H, and BROMEHEAD, C E N. 1915. Geology of the country around Windsor and Chertsey. *Memoir of the Geological Survey of Great Britain.*

DINES, H G, and EDMUNDS, F H. 1929. The geology of the country around Aldershot and Guildford. *Memoir of the Geological Survey of Great Britain.*

DRISCOLL, R. 1983. The influence of vegetation on swelling and shrinking of clay soils in Britain. *Geotechnique*, Vol. 33, 93–105.

EDMUNDS, F H. 1927. The Chalk at Guildford (correspondence). *Geological Magazine*, Vol. 64, 95.

EDMUNDS, W M, KINNIBURGH, D G, and MOSS, P D. 1992. Trace metals in interstitial groundwaters from sandstones: acidic inputs to shallow groundwaters. *Environmental Pollution*, Vol. 77, 129–141.

EDWARDS, R A, and FRESHNEY, E C. 1987. Lithostratigraphical classification of the

Hampshire Basin Palaeogene deposits (Reading Formation to Headon Formation). *Tertiary Research*, Vol. 8, 43–73.

ELLISON, R A. 1983. Facies distribution in the Woolwich and Reading Beds of the London Basin, England. *Proceedings of the Geologists' Asociation*, Vol. 94, 311–319.

ELLISON, R A, KNOX, R W O'B, JOLLEY, D W, and KING, C. 1994. A revision of the lithostratigraphical classification of the early Palaeogene strata of the London Basin and East Anglia. *Proceedings of the Geologists' Association*, Vol. 105, 187–197.

ELLISON, R A, and WILLIAMSON, I T. 1999. Geology of the Windsor and Bracknell district — a brief explanation of the geological map. *Sheet Explanation of the British Geological Survey*, 1:50 000 Sheet 269 Windsor (England and Wales).

FLATT, A G, HEARSUM, P G, and others. 1976. Records of wells around Aldershot: inventory for one-inch geological Sheet 285, new series. Well Inventory Series [Metric Units]. *Institute of Geological Sciences*, NERC., HMSO. 123pp.

FORSTER, A. CULSHAW, M G. and BELL, F G. 1995. Regional distribution of sulphate in rocks and soils of Britain. 95–104 *in* Engineering geology of construction. EDDLESTON, M, WALTHALL, S, CRIPPS, J C, and CULSHAW, M G (editors). *Geological Society Engineering Geology Special Publication*, No. 10.

GIBBARD, P L. 1982. Terrace stratigraphy and drainage history of the Plateau Gravels of north Surrey, south Berkshire and north Hampshire, England. *Proceedings of the Geologists' Association*, Vol. 93, 369–384.

GIBBARD, P L. 1985. *The Pleistocene history of the Middle Thames Valley.* (Cambridge: Cambridge University Press.)

HESTER, S W. 1965. Stratigraphy and palaeogeography of the Woolwich and Reading Beds. *Bulletin of the Geological Survey of Great Britain*, No. 23, 117–137.

IRVING, A. 1885. General section of the Bagshot strata from Aldershot to Wokingham. *Quarterly Journal of the Geological Society of London*, Vol. 41, 492–510.

JOHN, D T. 1980. The soils and superficial deposits of the North Downs of Surrey. 101–130 *in* The shaping of Southern England. JONES, D K C (editor). *Institute of British Geographers Special Publication*, No. 11.

KING, C. 1981. *The stratigraphy of the London Clay and associated deposits.* (Rotterdam: W Backhuys.)

KIRKALDY, J F. 1933. The Sandgate Beds of the Western Weald. *Proceedings of the Geologists' Association*, Vol. 34, 270–311.

KNIGHT, R I. 1999. Phosphates and phosphogenesis in the Gault Clay (Albian) of the Anglo–Paris Basin. *Cretaceous Research*, Vol. 20, 507–521.

LAKE, R D, and SHEPHARD-THORN, E R. 1985. The stratigraphy and geological structure of the Hog's Back, Surrey and adjoining areas. *Proceedings of the Geologists' Association*, Vol. 96, 7–21.

LAKE, R D, and SHEPHARD-THORN, E R. 1991. Geology of the area between Farnham and Guildford, Surrey. *British Geological Survey Technical Report*, WA/91/33.

LYONS, H G. 1887. On the London Clay and Bagshot Beds of Aldershot. *Quarterly Journal of the Geological Society of London*, Vol. 43, 431–442.

MATHERS, S J. 2000. Geology of the Reading district — a brief explanation of the geological map. *Sheet Explanation of the British Geological Survey*, 1:50 000 Sheet 268 Reading (England and Wales).

MEYER, C J A. 1868. On the Lower Greensand of Godalming. *Proceedings of the Geologists' Association*, Volume Supplement to Volume 1, 1–20.

MONCKTON, H W, and HERRIES, R S. 1886. The Bagshot Beds of the London Basin. *Quarterly Journal of the Geological Society of London*, Vol. 42, 402–417.

OWEN, H G. 1975. The stratigraphy of the Gault and Upper Greensand of the Weald. *Proceedings of the Geologists' Association*, Vol. 86, 475–498.

OWEN, H G. 1984. The Albian Stage: European province chronology and ammonite zonation. *Cretaceous Research*, Vol. 5, 329–344.

OWEN, H G. 1992. The Gault–Lower Greensand Junction Beds in the northern Weald (England) and Wissant (France), and their depositional environment. *Proceedings of the Geologists' Association*, Vol. 103, 83–110.

PURTON, L M A, and BRASIER, M D. 1999. Giant protist *Nummulites* and its Eocene environment: life span and habitat insights from $\delta^{18}O$ and $\delta^{13}C$ data from *Nummulites* and *Venericardia*, Hampshire basin, UK. *Geology*, Vol. 27, 711–714.

PRESTWICH, J. 1847. On the main points of structure and the probable age of the Bagshot Sands. *Quarterly Journal of the Geological Society of London*, Vol. 3, 378–409.

ROBINSON, N D. 1986. Lithostratigraphy of the Chalk Group of the North Downs, southeast England. *Proceedings of the Geologists' Association*, Vol. 97, 141–170.

RUFFELL, A H. 1992a. Early to mid-Cretaceous tectonics and unconformities of the Wessex Basin (southern England). *Journal of the Geological Society of London*, Vol. 149, 443–454.

RUFFELL, A H. 1992b. Correlation of the Hythe Beds Formation (Lower Greensand Group: early–mid-Aptian), southern England. *Proceedings of the Geologists' Association*, Vol. 103, 273–291.

SHEPHARD-THORN, E R, and LAKE, R D. 1991. Geology of the area between Albury and Dorking, Surrey. *British Geological Survey Technical Report*, WA/91/34.

SIMPSON, M I. 1985. The stratigraphy of the Atherfield Clay Formation (Lower Aptian: Lower Cretaceous) at the type and other localities in southern England. *Proceedings of the Geologists' Association*, Vol. 96, 23–45.

SMALLEY, S, and WESTBROOK, G K. 1982. Geophysical evidence concerning the southern boundary of the London Platform beneath the Hog's Back, Surrey. *Journal of the Geological Society of London*, Vol. 139, 139–146.

THURRELL, R G, WORSSAM, B C, and EDMONDS, E A. 1968. Geology of the country around Haslemere (Explanation of Geological Sheet 301). *Memoir of the Geological Survey of Great Britain.*

WOODS, M A. 1997. A review of the stratigraphy of the Chalk Group of the Reading (268), Aldershot (285) and Reigate (286) districts. *British Geological Survey Technical Report (Stratigraphy Series)*, WH/97/99R.

WOODS, M A. 1998. A stratigraphical review of the Cretaceous formations of the Aldershot Sheet (285). *British Geological Survey Technical Report (Stratigraphy Series)*, WH/98/156R.

WORSSAM, B C. 1973. A new look at river capture and the denudation history of the Weald. *Report of the Institute of Geological Sciences*, No. 73/17.

WORSSAM, B C. 1978. The stratigraphy of the Weald Clay. *Report of the Institute of Geological Sciences*, No. 78/11.